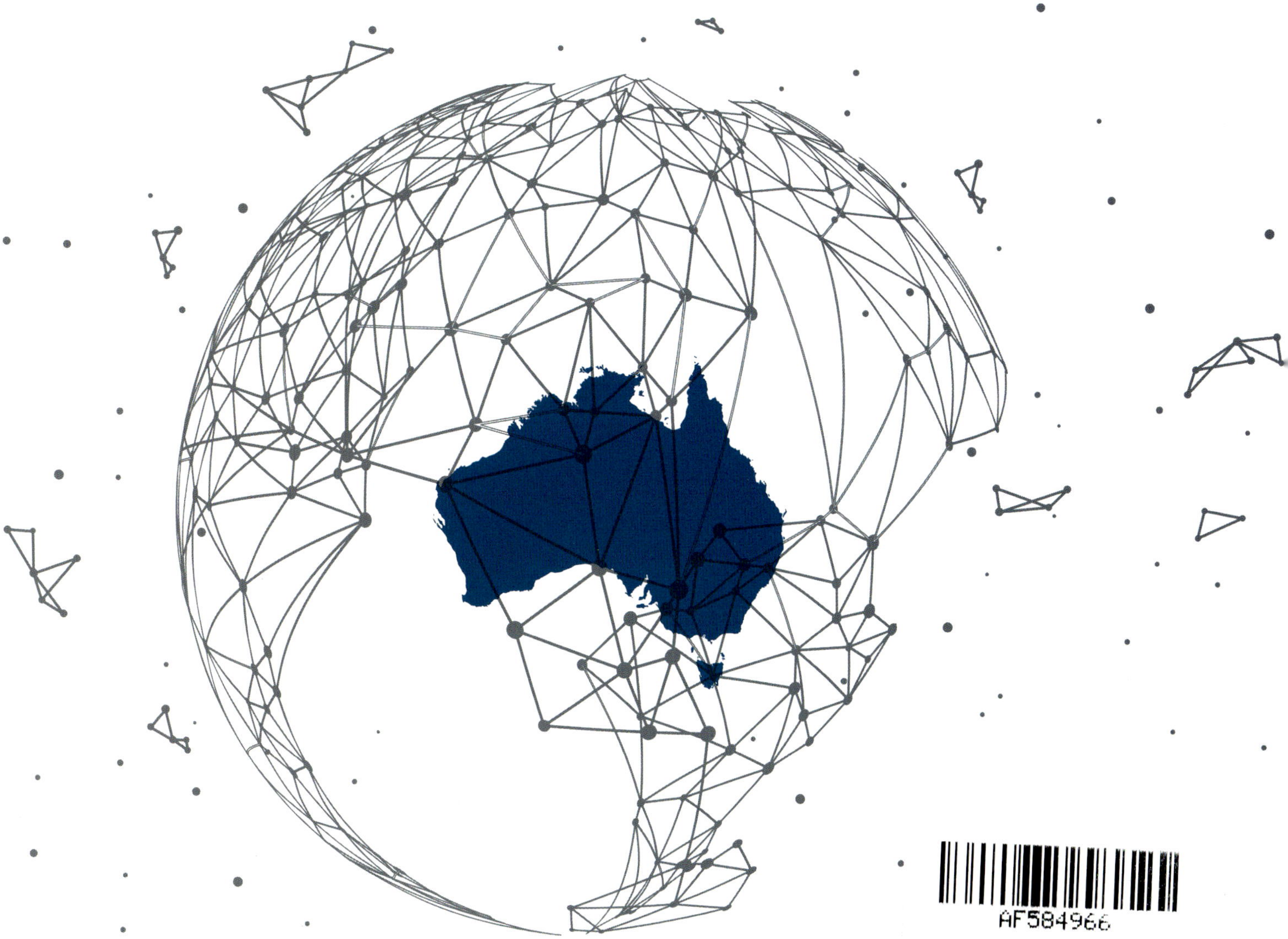

# NATURAL DISASTERS

AF584966

First Published 2025 by
Redback Publishing
Suite 6, 13a Narabang Way,
Belrose NSW 2085
Australia

www.redbackpublishing.com.au
orders@redbackpublishing.com



ISBN 978-1-761401-61-9 PBK



Author: Peter Turner
Editor: Caroline Thomas
Designer: Redback Publishing

Original illustrations © Redback Publishing 2025
Originated by Redback Publishing

Acknowledgements
Abbreviations: l—left, r—right, b—bottom, t—top, c—centre, m—middle
We would like to thank the following for permission to reproduce photographs: (Images © shutterstock),

p8b Christchurch, New Zealand - March 12: A brick house on historic Cranmer Square collapses from the impact of the massive earthquake on March 12, 2011 in Christchurch, by Nigel Spiers, via Shutterstock.com, p22bl Photograph of HMAS Arrow beached in Francis Bay in early 1975, by Billbeee, [CC BY-SA 3.0 (http://creativecommons.org/licenses/ by-sa/3.0/)] via Wikimedia Commons, p22br Damaged houses after the passage of Cyclone Tracy on Christmas day 1974 in Darwin, Australia, by Billbeee, [CC BY-SA 3.0 (http://creativecommons.org/licenses/ by-sa/3.0/)] via Wikimedia Commons.

Every effort has been made to contact copyright holders of any material reproduced in this book. Any omissions will be rectified in subsequent printings if notice is given to the publisher.

A catalogue record for this book is available from the National Library of Australia

# CONTENTS

# THE NATURAL RHYTHM

A natural disaster is a devastating event that is nonetheless a part of the natural rhythms and forces that shape our planet. Natural disasters cause great destruction, and threaten the lives of people and animals. They can affect whole ecosystems, and throughout the history of Earth they have shaped the way it looks and have determined the success or failure of the species that inhabit it.

## CLIMATE CHANGE

Scientists now know that the Earth is experiencing more severe weather events because of human activity. Manufacturing industries and our environmental practices have caused excessive greenhouse gases to enter the Earth's atmosphere. This has led to global warming which has caused our climate to change. The consequences of climate change are far-reaching and include the increasing severity and frequency of floods, droughts and storms.

## ADAPTATIONS AND NATURAL DISASTERS

Some species have adapted to the environment and now rely on natural disasters for survival. Some Australian plants, for example, need the fierce heat and smoke of a bushfire in order to germinate and regenerate. Some trees need regular flooding to keep them healthy and to encourage them to flower.

# SHAPING A NATION

Over time, Australia has experienced many natural events that have helped to create the land we recognise today. For example, the dry, desert inland was once a vast sea, teeming with life. Today, that same area supports life of a very different kind. In modern Australia, these natural events play an unpredictable and sometimes disastrous part in all our lives. Drought, fire, tropical cyclones, floods and earthquakes affect all Australians, either directly or indirectly. We might suffer personal tragedy as a result of a natural disaster, or we might be affected economically as the food we take for granted suddenly becomes scarce and expensive.

!

Natural disasters remind us that we live in the natural world, that we are a part of it and that we are subject to its forces. Understanding this helps us to prepare, to protect and to recover.

In Australia, natural disasters have become an important part of our social history. Some events, such as Cyclone Tracy, the Black Saturday bushfires and the Newcastle earthquake, are remembered as much for the emotional effect they had on our nation as they are for the destruction they caused. Disasters bring people together because they are a common enemy. They show us at our most vulnerable and at our strongest, kindest and most united. They shape our landscape, and they help to shape our sense of ourselves as a nation.

GIVE WAY

# TYPES OF NATURAL DISASTERS

## TORNADO

A vortex of wind in the shape of a tunnel that can cause a lot of damage

## HAIL

Chunks of ice that fall during a storm or rain

## TSUNAMI

Huge waves caused by an earthquake or volcanic eruption

## LANDSLIDE

When a mountain or cliff face collapses

## BUSHFIRE

A fire in grass, bush or woodland that is hard to control

## EARTHQUAKE

When the Earth's crust moves and shakes violently

## CYCLONES

Large masses of rotating winds with an area of low pressure at the centre. Wind speeds are extreme and destructive

## BLIZZARD

Severe snowstorm with heavy winds

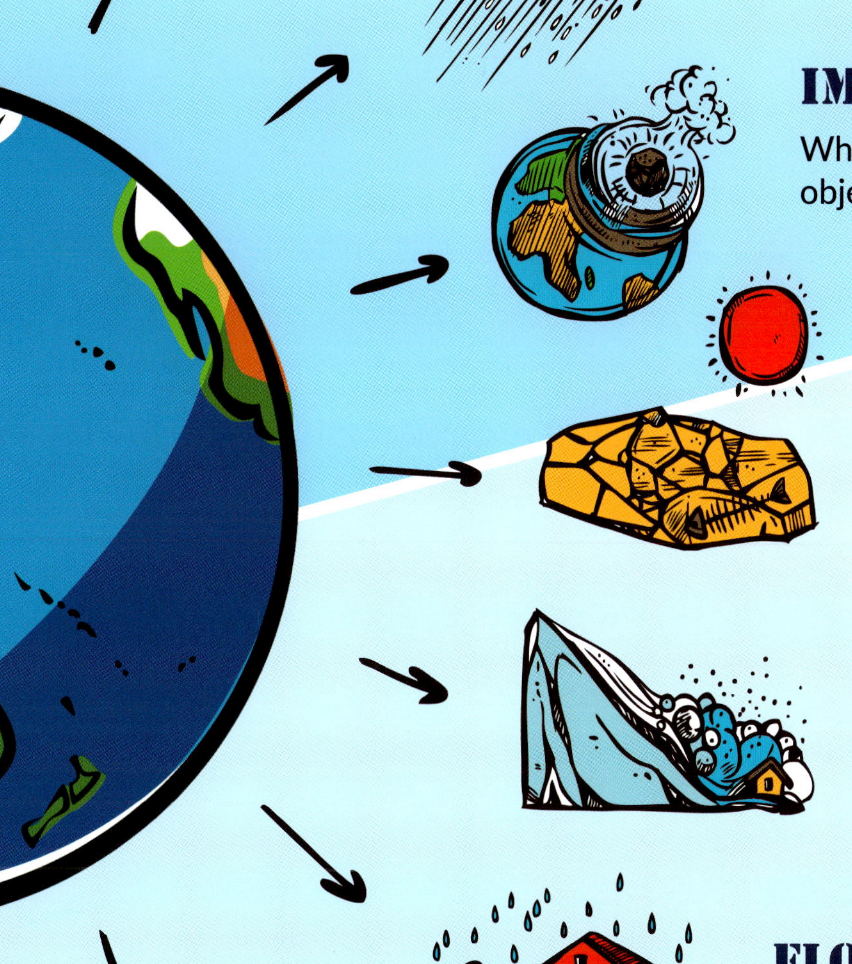

## IMPACT EVENT

When an astronomical object collides with Earth

## DROUGHT

A prolonged period of abnormally low rainfall

## AVALANCHE

A large amount of snow that slides down the side of a mountain

## FLOOD

When water overflows beyond its normal limits and dry land is inundated

## VOLCANIC ERUPTION

When a volcano discharges steam and volcanic materials, including rocks and lava

# MANAGING NATURAL DISASTERS

A community that is well prepared for a natural disaster is more likely to be well equipped to deal with one when it occurs. A well prepared community can also cope better in the aftermath of the disaster. This is when the hard work begins for cleanup, rebuilding and replanting, and the social and emotional recovery of the community members.

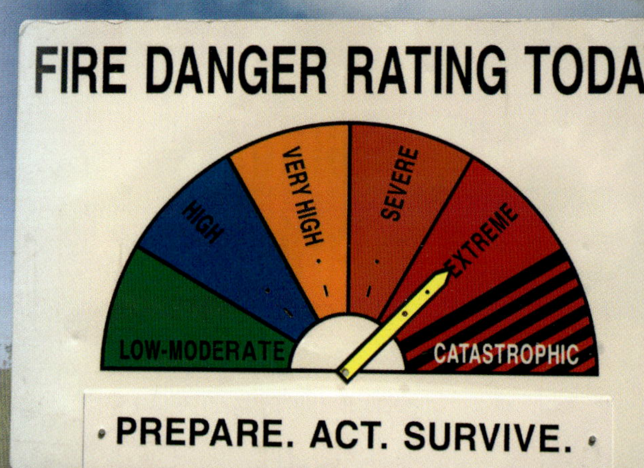


## EARTHQUAKES

Earthquakes happen so quickly that people make the mistake of rushing outside, where they become vulnerable to falling powerlines, crashing debris and flying glass. If you are indoors, stay there and shelter away from windows, under a doorframe or a table.

If you are outside, move away from powerlines, buildings, walls and trees. Wherever you are, look up. If you are under a fan, ceiling light, shelf or any structure that might shake loose, move away.

# DROUGHT

The Australian Bureau of Meteorology, along with the World Meteorological Organization, monitors weather patterns and contributes to the Climate Information and Prediction Services (CLIPS). This service aims to give advanced warning of potential droughts.

It is not possible to make Australia drought-proof, but there are water-saving measures that can help to reduce the demand on reservoirs. Governments introduce water restrictions in times of drought to help households to reduce their usage.

In rural areas, planting trees can help to stop the topsoil from blowing away, and covering irrigation channels can reduce evaporation.

!

Australians experience drought regularly, and should always be conscious of how they use water.

Having shorter showers, using efficient showerheads, recycling greywater, installing rainwater tanks in gardens and fixing dripping taps all help.

## FLOODS

Sometimes, warnings can be issued before a flood happens. Warnings give people time to move to higher ground and to prepare their homes for flooding.

If there is any danger of potential flooding in your area, it is important to have an evacuation plan.

- Turn gas and electricity off at the point where they enter the property.
- Valuable items should be placed up high.
- Open freezer and refrigerator doors so that they cannot trap air. Floating objects cause damage to other items and people.
- Keep a portable radio with you to stay up to date on safety advice. There may be no electricity available during a flood.
- Never attempt to swim or drive through flood waters.
- After a flood, all drinking water should be boiled until the water supply is made safe.

## BUSHFIRES

Digital Earth Australia Hotspots is a national bushfire monitoring system. This internet-based satellite tracking system allows people to see where a fire has started and where it is likely to spread. Firefighting organisations use DEA Hotspots to locate and manage fires.

Governments enforce total fire bans on days when conditions are very hot and dry. On total fire ban days, it is illegal to light an uncontained fire outside, or use equipment that produces sparks.

!

People who live in or near the bush, and in towns, should ensure their gardens and gutters are free of combustible litter, such as leaves and twigs. They should also have a fire plan and be well prepared with access to a non-mains fed water supply - such as a pool or dam – if they intend to stay and defend their property.

# DROUGHT

In Australia, drought is a natural consequence of the continent's geographical position. A drought occurs when an unusually long period of low rainfall results in there not being enough water to meet the normal needs of agriculture and of the community. Since Australia is the driest inhabited continent in the world, there is more to a drought than simply dry conditions. In many parts of Australia, low rainfall is perfectly normal. These places are not in a constant state of drought. They are simply experiencing the natural conditions of their climate.

## THE EFFECTS OF DROUGHT

Drought affects everyone and everything, especially farmers. Australia's economy is badly affected by a drought. Other problems include:

- Plants dying
- Rivers and creeks drying up
- Wildlife threatened and breeding cycles interrupted
- Dry, fertile soil being blown away
- Crops failing
- Farm animals being unable to graze
- Farmers suffering financial hardship
- Food costs increasing for everyone
- Water restrictions coming into force
- Bushfire risk increasing

## SUBTROPICAL HIGH

Periods of drought are a normal part of Australia's climate. This is because the continent is situated at a latitude that is strongly influenced by the subtropical high pressure belt.

The subtropical high exists in both the Southern and Northern Hemispheres. In these regions, just north and just south of the Equator, the air is warm and dry, skies are often clear and there is low rainfall.

## EL NIÑO

El Niño means 'the boy' in Spanish. It is a warm ocean current that moves through the Pacific Ocean every three to eight years. As the current moves, the air pressure above it drops, diverting rain-bearing winds away from land. This weather pattern is known as the El Niño Southern Oscillation. It occurs in the Pacific region but can effect the weather in many parts of the world.

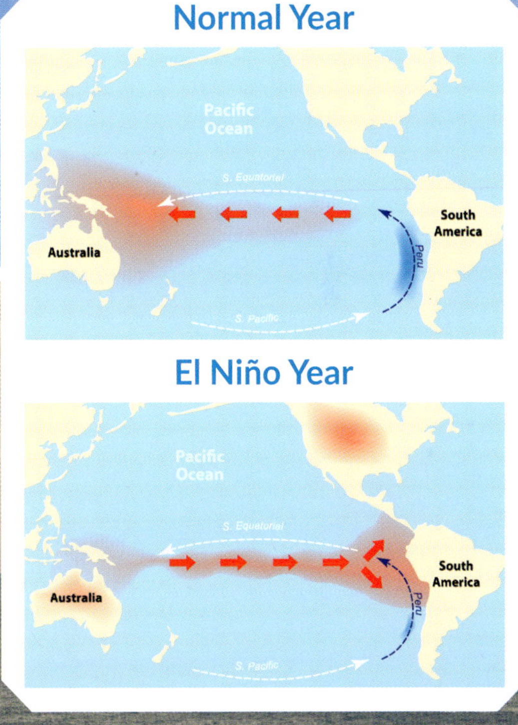

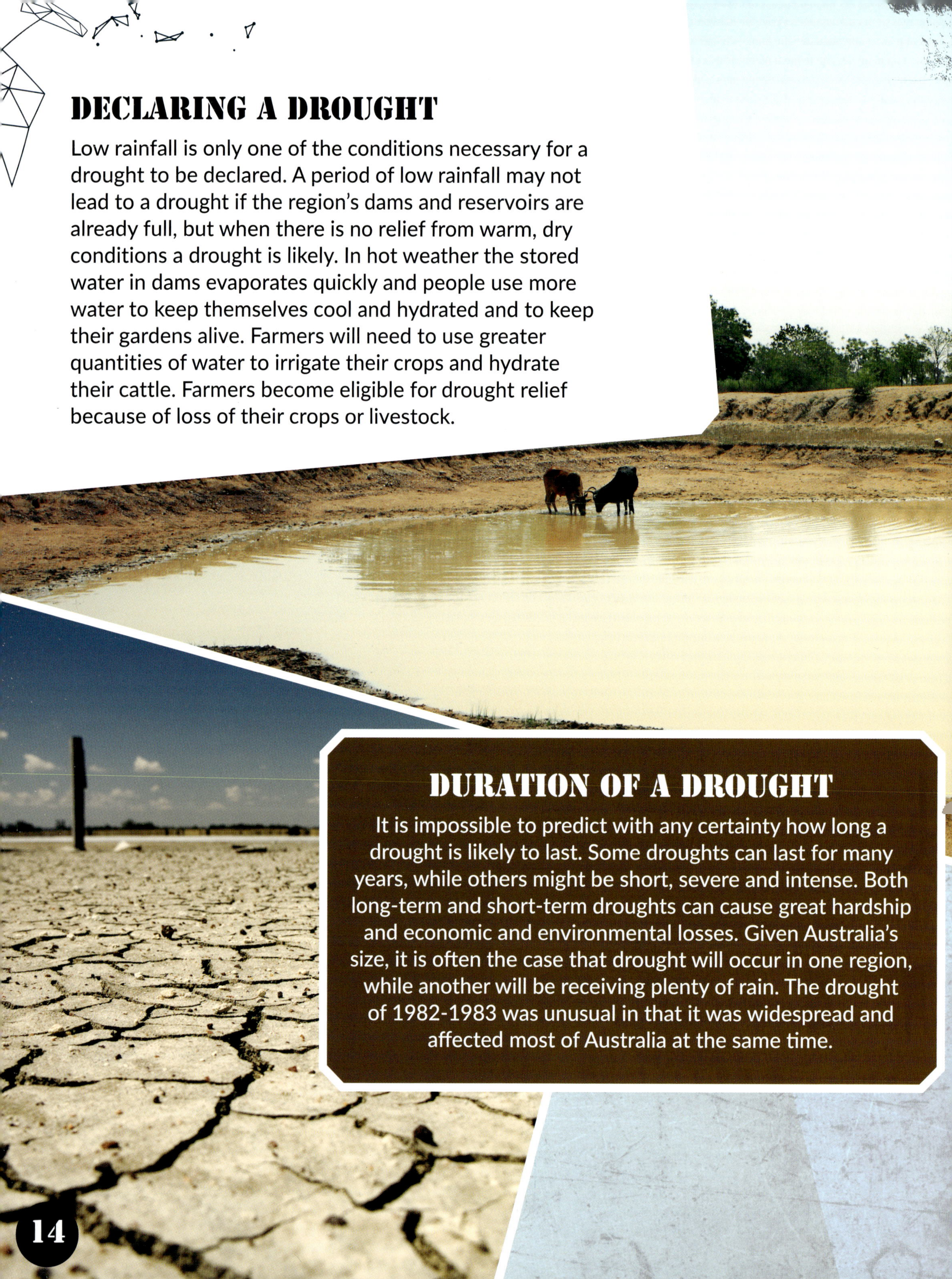

## DECLARING A DROUGHT

Low rainfall is only one of the conditions necessary for a drought to be declared. A period of low rainfall may not lead to a drought if the region's dams and reservoirs are already full, but when there is no relief from warm, dry conditions a drought is likely. In hot weather the stored water in dams evaporates quickly and people use more water to keep themselves cool and hydrated and to keep their gardens alive. Farmers will need to use greater quantities of water to irrigate their crops and hydrate their cattle. Farmers become eligible for drought relief because of loss of their crops or livestock.

## DURATION OF A DROUGHT

It is impossible to predict with any certainty how long a drought is likely to last. Some droughts can last for many years, while others might be short, severe and intense. Both long-term and short-term droughts can cause great hardship and economic and environmental losses. Given Australia's size, it is often the case that drought will occur in one region, while another will be receiving plenty of rain. The drought of 1982-1983 was unusual in that it was widespread and affected most of Australia at the same time.

# 2009 EASTERN AUSTRALIAN DUST STORM

In 2009, Australia experienced its worst dust storm in 70 years. A huge dust storm swept across New South Wales and Queensland over two days in September 2009, affecting Canberra, Sydney and Brisbane.

The orange cloud was visible from space and NASA measured the distance of its plume from the northern edge to its southern edge to be 3,450 km. The Australian Bureau of Meteorology described the event as "pretty incredible". When the dust blocked the sun it caused a temperature drop on the ground, much like what scientists predict may follow a nuclear war. This nuclear winter is predicted to occur when smoke or dust blocks the sun for a prolonged period of time.

!

The CSIRO estimated that the storm swept 16 million tonnes of dust from the deserts of Central Australia and dumped them in Sydney Harbour and the Tasman Sea.

# BUSHFIRES

Bushfires are not always disastrous and they play a vital role in the healthy life cycle of forest ecosystems. Many Australian native trees and shrubs depend on fire and smoke to create the right conditions for their seeds to germinate.

The Australian bush generally recovers well after fire. Unless a bushfire is particularly intense or is moving at great speed, native Australian animals can protect themselves by burrowing, or outrunning the flames. A bushfire becomes a natural disaster when it sweeps through areas that are inhabited by people, destroying property and crops, and taking lives.

## CAUSES OF BUSHFIRES

Bushfires need the right atmospheric conditions and the right fuel to take hold and rage out of control. The dominant tree in the Australian bush is the eucalypt, and its leaves have a high level of oil. This makes them very flammable, especially when there is little moisture in the air.

The Australian bush becomes vulnerable to fire after dry weather, when large quantities of fuel such as dry leaves, twigs, sticks, bark and dead bushes litter the forest floor.

# HOW BUSHFIRES AFFECT AUSTRALIA

The most devastating effects of bushfire are loss of life and destruction of people's homes. Fierce bushfires leave nothing in their wake. They are terrifying to experience. If the conditions are right, they can explode upon people with astonishing suddenness.

In high winds, embers from a fire can blow well ahead of the flames and set an area alight that is some distance away from the main fire. Firefighters understand how bushfires move and know how to fight fire with fire. Deliberately burning an area along the predicted path of a wildfire removes the fuel source and stops the fire spreading. This can only be done in low wind. Fierce winds can blow fires in all directions and the fires even create their own winds.

## CROWNING

In particularly fierce bushfires, the fire moves at great speed by leaping from the burning crown, or top, of one tree to another. This is called crowning, and is the most dangerous type of bushfire. These fires are bombed from the air, using aircraft to drop fire suppressants or water to douse them.

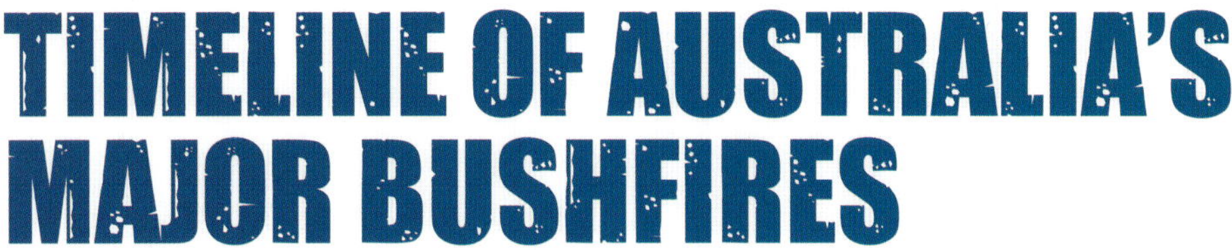

# TIMELINE OF AUSTRALIA'S MAJOR BUSHFIRES

There are hundreds of bushfires in Australia each year. Some of them occur in remote areas, and some of them are much closer to where people live. From time to time, bushfires have been so extreme that they have had a disastrous effect on the lives of many people, and have caused the whole nation to feel some of the horror that comes from such tragedies.

## JANUARY TO MARCH 1926

Over 1,000 properties in Victoria were either destroyed or damaged and 60 people lost their lives in a series of major bushfires. The most devastating fire was on February 14, which later became known as Black Sunday.

## 13TH JANUARY 1939

The Black Friday bushfires in Victoria were among the worst natural wildfires in the world. Almost 20,000 square kilometres of land burned and several towns were destroyed. Seventy-one people lost their lives.

## THE SUMMER 1943-44

1943–44 was the driest summer ever recorded in Melbourne and numerous bushfires devastated areas of Victoria during this period. A year later, the Country Fire Authority was established to co-ordinate rural fire brigades.

## 17TH FEBRUARY 1967

One hundred and ten separate fire fronts raged through southern Tasmania taking 62 lives and injuring hundreds of people. Over 7,000 people lost their homes and an estimated 62,000 farm animals were killed.

## JANUARY 1969

Once again parts of Victoria were devastated by bushfire. Twenty-three people lost their lives and 230 homes were destroyed.

## ASH WEDNESDAY 1983

On 16 February, a series of fires caused widespread destruction across South Australia and Victoria. In the Dandenong Ranges, the bushfire surrounded 17 firefighters, all of whom lost their lives. There were 75 lives lost altogether.

## LINTON BUSHFIRE 1998

Five volunteer firefighters died in the 1998 Linton bushfire. Their deaths led to changes in safety operating procedures in South Australia's Country Fire Service and the Victorian Country Fire Authority.

## CANBERRA, 18TH JANUARY 2003

Fires in Kosciuszko National Park and Namadgi National Park spread into Canberra's suburbs. Four people lost their lives, more than 530 homes were destroyed and 99% of the Namadgi National Park was destroyed.

## VICTORIAN BUSHFIRES 2005

Four people lost their lives. Hundreds of homes and farm buildings were lost and over 65,000 farm animals perished in the fires.

## BLACK SATURDAY BUSHFIRES 2009

The Black Saturday fires began around 7 February 2009 and resulted in Australia's highest ever loss of life from a bushfire. 180 people died, another 400 were injured, and over 2,029 homes were destroyed.

The week before the fires, Melbourne broke records with three consecutive days above 43°C. By the Saturday (soon to become known as Black Saturday) winds had reached speeds in excess of 100 kilometres per hour and a top temperature of 46.4°C was recorded in Melbourne. Over 400 fires burned on Black Saturday.

## SUMMER OF FIRE 2019/2020

The Australian Government declared a State of Emergency when bushfires destroyed 19 million hectares of land in southeastern Australia between August 1, 2019 and February 13, 2020, before 400 millimetres of rain fell in one day, flooding out the last of the fires. An estimated one billion animals, not including birds, reptiles, fish and insects were killed. American and Canadian firefighters were called in to help battle the fires and New Zealand, Singapore and Papua New Guinea offered military support. Over 6,500 buildings were destroyed and 33 people lost their lives including three American firefighters. The event triggered a global response, with one social media campaign, from comedian Celeste Barber, raising over $52 million. Many other fundraising events and wildlife support groups were organised. 2019 was Australia's driest year since records began in 1900 and the annual mean temperature was 1.52°C above average. Changes to our climate mean more intense fire seasons will continue to occur and governments are now working hard to find long-term solutions to tackle future disasters of this kind.

# TROPICAL CYCLONES

Tropical cyclones – also called hurricanes or typhoons – are strong storms that develop over the ocean in regions that are close to the tropics. They are different from other storms in that they spin, gathering speed . A satellite photograph shows that a cyclone looks like a swirling mass of cloud, with a central point, called the eye. The word 'cyclone' comes from a Greek word meaning 'wheel' or 'coil of a snake'.

## CYCLONE RISK

Cyclones are at their most powerful out over the ocean and close to the coast. As they move inland, they begin to lose energy. They bring with them flooding rains, destructive winds and storm surges that can threaten coastal communities with sudden rises in the sea level.

## THE AUSTRALIAN CYCLONE SEVERITY SCALE

This scale was adopted by the Australian Bureau of Meteorology in 1992 and is categorised numerically.

| CAT 1 | CAT 2 | CAT 3 | CAT 4 | CAT 5 |
|---|---|---|---|---|
| <125 km/h | 125-164 km/h | 165-224 km/h | 225-279 km/h | >280 km/h |
| Minimal Damage | Moderate Damage | Extensive Damage | Extreme Damage | Catastrophic Damage |

!

Satellites and weather-tracking devices have made it possible to issue warnings ahead of a cyclone's development and to see where it is at a particular time. This is especially useful for the shipping industry. Cyclones move in unpredictable directions. At one moment they might appear to be passing safely offshore, while at the next they might change direction and head inland.

# TROPICAL CYCLONE CASE STUDIES

Communities in Australia have regularly endured the destructive force of tropical cyclones. In 1899, 300 people died during a cyclone at Bathurst Bay, in northern Queensland. Before the advent of weather-tracking systems and good communications equipment, ships out at sea and coastal communities were vulnerable. Even with modern satellites and communications systems, cyclones can still have devastating consequences.

## CYCLONE TRACY, 1974

On Christmas morning in 1974, Cyclone Tracy swept through Darwin, destroying most buildings, killing 49 people in the town and 16 at sea.

Cyclone Tracy was not a particularly large cyclone, but the winds it generated were very strong. Tracy was slow-moving, with winds that blew through Darwin for more than five hours. Darwin's buildings were not constructed to withstand such forces, and most of them collapsed. When the rest of Australia saw images of Darwin after Cyclone Tracy the most common response was that it looked as if a large bomb had been dropped on the town.

Cyclone Tracy was such a catastrophe that most of Darwin's population had to be evacuated. There was no food, no power and no drinkable water. Some people drove south to the towns of Adelaide River, Tennant Creek and Katherine. Thousands of others were airlifted in both civilian and military aircraft to cities in the south. After Cyclone Tracy, Darwin has now been rebuilt with buildings that should withstand cyclones.

## CYCLONE YASI, 2011

Early on 3 February 2011, Tropical Cyclone Yasi made landfall in far north Queensland, bringing 285 kilometre per hour winds. Yasi was the costliest cyclone to ever hit Australia, with damage exceeding $3.5 billion.

## CYCLONE LARRY, 2006

On the morning of 20 March 2006, Cyclone Larry struck the north Queensland coast, near Innisfail. People knew Larry was coming and they were able to prepare and protect themselves. However, they could not protect Queensland's largest crop - bananas. What turned Larry from a cyclone into a natural disaster was the destruction of more than 90% of Australia's banana crop. Australians eat more than 15 million bananas each week, but for nine months after Cyclone Larry, the supply of bananas crashed. Australia does not import bananas because of the threat of disease, so the small number available for sale came from New South Wales, the Northern Territory and Western Australia. Bananas became very expensive. People in the industry were laid off work and crop losses were estimated to be around $350 million.

# FLOODS

Floods are an important part of many ecosystems. In Australia, the areas that border the Murray River and the Macquarie Marshes in New South Wales need to be regularly flooded in order to keep them healthy. Floods become natural disasters when they are so severe that they cause environmental damage, destroy crops and stock, damage buildings, roads and bridges, and take lives.

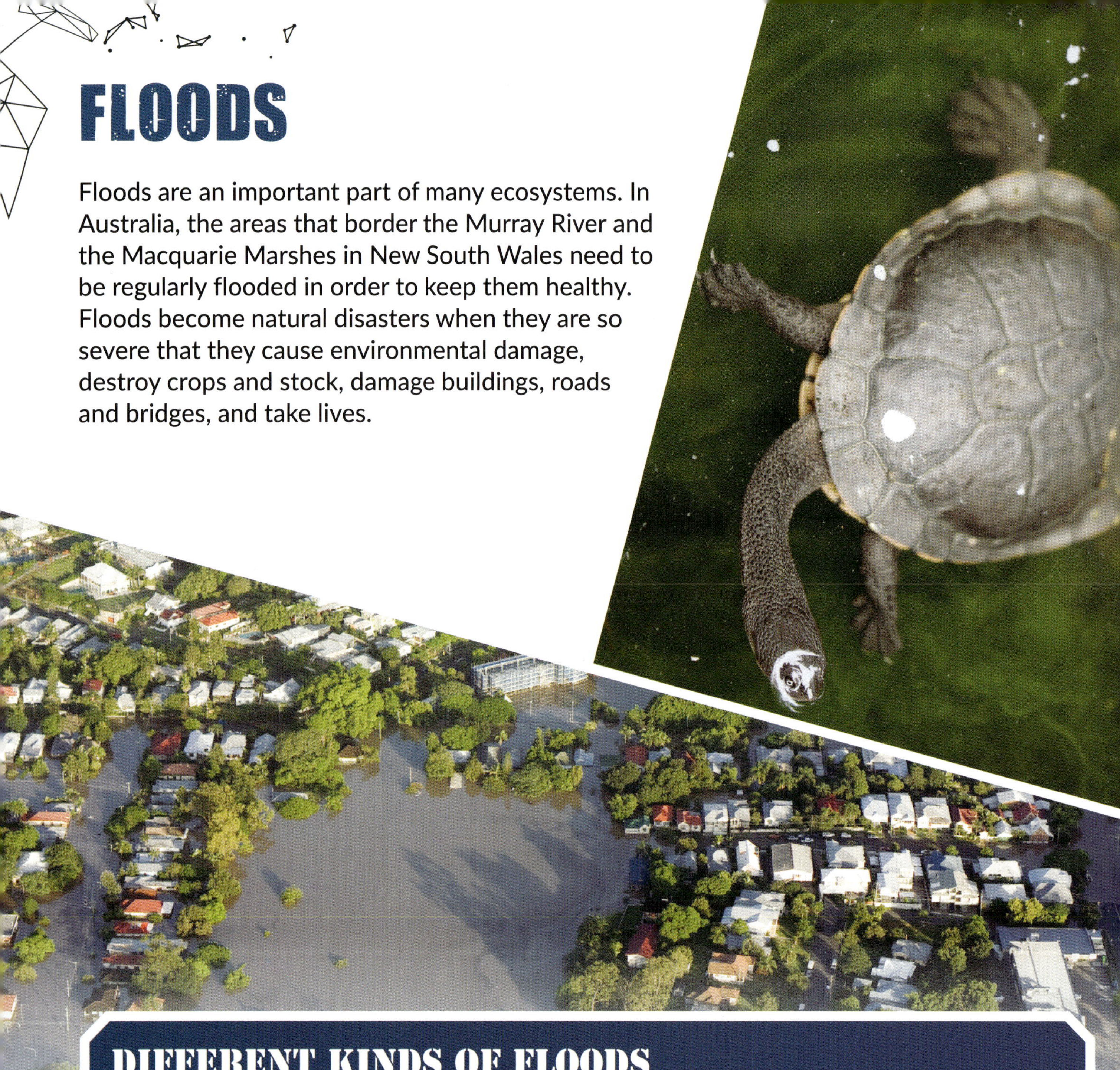

## DIFFERENT KINDS OF FLOODS

Floods occur in all parts of Australia, from the arid interior to the wet tropics. In northern regions of the country, most floods occur between November and May as a result of cyclones. In southern areas, floods occur more often in winter and spring.

Floods happen when heavy rainfall causes creeks and rivers to become swollen with large volumes of water. A flood occurs when this water breaks the banks and spreads out across the land. A flood may occur many days after heavy rainfall upstream.

## STORM DAMAGE

One hailstorm in Sydney on 14 April 1999 caused more than $1.7 billion damage. The Australian Bureau of Meteorology estimates that floods cost the community more than $400 million each year.

## LA NIÑA

La Niña means 'the girl' in Spanish. It is a weather pattern that brings higher than average rainfall every three to five years. Known as La Niña Southern Oscillation, it has the opposite effect to El Niño Southern Oscillation. Floods are more likely to occur during La Niña periods. El Niño brings periods of drought.

## FLASH FLOOD ALERT!

One of the most destructive kinds of flood is a flash flood. Flash floods are the result of sudden, very heavy rain. They are especially destructive in cities where drainage systems fail, and streets and houses are threatened with rushing water. The force of this water can be extraordinary. It is not unusual for cars and trucks to be swept away.

# AUSTRALIA'S WORST FLOODS

Damaging localised flooding is a regular occurrence in most parts of Australia. Some floods are so devastating that their effects are felt far and wide, and are remembered for many years afterwards.

## THE MAITLAND FLOOD, FEBRUARY 1955

Between 1954 and 1956, Australia experienced a La Niña-influenced event. Heavy rain fell over eastern Australia and, on 23 February 1955, the Hunter River in New South Wales burst its banks and roared into the town of Maitland, in the Hunter Valley.

The water spread over an area almost twice the size of Tasmania. 25 people lost their lives and photographs and film footage of the devastation shocked Australians everywhere. The damage has been estimated at more than $1.3 billion. The flood of 1955 remains the largest flood to affect the Hunter Valley.

## KATHERINE FLOODS, JANUARY 1998

On 25 January 1998, rain began falling in Katherine in the Northern Territory, as a result of tropical Cyclone Les which had begun the previous day over the Gulf of Carpentaria. The rain was heavy and fell continuously for two days.

The region's average January rainfall was 236 millimetres, but over 900 millimetres of rain fell in January 1998. The Katherine, Roper and Daly Rivers were already full when they took on a further 420 millimetres of rain in 48 hours.

Katherine had begun to flood even before the river rose. Stormwater drains burst and, instead of carrying water away, spilled it into the streets.

When the Katherine and Daly Rivers finally broke their banks, more than 1,000 square kilometres of land were flooded.

The whole of the central business district was flooded, with water in the main street rising to a depth of two metres. Houses and businesses were devastated and more than 5,000 people were forced to evacuate. 1,200 homes and 500 businesses were flooded.

As well as dealing with disease, food shortages and fast-flowing objects, Katherine's rescuers also had to be wary of snakes and crocodiles. Three people died because of the floods and the damage was estimated at $70 million.

## QUEENSLAND FLOODS, 2010–2011

Beginning in December 2010 and continuing into 2011, a series of floods devastated parts of Queensland. Flooding was so bad that torrents of water flowed into the Lockyer Valley, in what was described as an inland tsunami.

Thirty-five people died and approximately 200,000 people were affected across 90 towns, with many suburbs of Brisbane flooded. Three-quarters of Queensland was declared a disaster zone.

# EARTHQUAKES

In towns and cities, earthquakes can cause buildings to collapse, and there is a danger of being crushed, or hit by falling debris. Earthquakes cause landslides under which houses can be buried, and fires can start if gas lines are broken, or fuels or chemicals are spilled. Earthquakes that take place under the ocean can generate a tsunami, a rapidly moving, giant wave that can cause great destruction to coastal communities.

## HOW DO EARTHQUAKES HAPPEN?

The surface of the Earth is made up of large separate pieces of crust, called tectonic plates. They shift and move as they float on the Earth's magma layer. When they grind against or collide with neighbouring plates, they release seismic energy that moves in waves up towards the Earth's surface.

Seismic activity can occur anywhere on the Earth's surface, but is more frequent where the tectonic plates meet.

More than 80% of earthquakes occur along a region called the Pacific Ring of Fire. The countries most affected are New Zealand, Fiji, Papua New Guinea, the Philippines, Japan, the east coast of Russia, the west coast of Canada, and North and South America. Earthquakes that occur in places that sit in the middle of plates, rather than at the edges, are called intra-plate earthquakes.

## NEWCASTLE, 28 DECEMBER 1989

Newcastle is a large city located to the north of Sydney in New South Wales. At 10.27 am on 28 December 1989, an earthquake struck the city. It caused widespread destruction and claimed the lives of 13 people.

The Newcastle earthquake registered 5.6 on the Richter scale and was felt as far as 800 kilometres away from its epicentre. Thankfully, the earthquake struck at a time when the town was less populated than usual. School holidays meant that children were home, and a strike by bus drivers meant public transport had brought fewer people than usual into central Newcastle that day. Damage was estimated at more than $4 billion, and 160 people were injured. Three hundred buildings had to be demolished and tens of thousands of buildings were damaged.

### AFTERSHOCKS

After a large earthquake, the area continues to shift until it stabilises itself. This results in aftershocks, which are like echoes of the original earthquake. Aftershocks can cause as much damage as the actual earthquake. They can occur within hours, days, weeks or even months after an earthquake.

### MEASURING EARTHQUAKES

Earthquakes can be measured using a seismograph. It records vibrations so that the time between seismic waves can be measured. This can determine the place where the earthquake began, which is its epicentre. Earthquakes are measured on the Richter scale, which indicates their magnitude or size.

# AROUND THE WORLD

Globally, there have been countless natural events that have had devastating impacts on the environment and on people. The myths and tales of ancient civilisations all contain references to devastating geological events.

## CENTRAL CHINA FLOODS

The worst natural disaster ever recorded was the 1931 Central China Floods. An estimated 3.7 million people died from drowning, disease and starvation, and a further 50 million people were impacted.

## BOXING DAY TSUNAMI

On 26 December 2004, the third largest earthquake ever recorded struck in the Indian Ocean off Sumatra, Indonesia. It unleashed a devastating tsunami that killed 230,000 people across 14 countries and caused widespread destruction. Hardest hit was Indonesia where 168,000 people were killed.

## HAITI EARTHQUAKE

In 2010, a catastrophic earthquake hit Haiti. It was followed by weeks of aftershocks. Millions of people were impacted by the earthquake and death toll estimates ranged from 100,000 to 160,000.

# GLOSSARY

**arid region** a very dry area
**atmosphere** the gases that surround the Earth
**climate** the weather conditions in a particular area
**climate change** a change in the usual climate patterns
**debris** rubbish, the result of destruction
**drought relief** financial help from the government during a drought
**ecosystem** a natural system that supports life within it
**El Niño** an irregularly occurring variation in winds and sea surface temperatures
**embers** burning particles
**equator** an imaginary line that runs around the Earth and is equally distant from the North and South Poles
**evacuate** to leave an area in an emergency
**geographical** having to do with a place's location
**geologically** having to do with rocks
**germinate** to begin to grow
**global warming** an increase in the normal annual temperature range of the Earth
**species** a class of living things with similar characteristics
**subtropical** lying outside the tropical zone and between the Tropics of Cancer and Capricorn
**suppressant** a substance used to smother or keep down a fire
**tectonic plates** slowly moving plates that form the Earth's surface
**tropics** those areas on or near the Equator

# INDEX